CAREC
ENERGY STRATEGY 2030

Common Borders. Common Solutions. Common Energy Future.

NOVEMBER 2019

CAREC

Central Asia Regional Economic Cooperation Program

ADB

ISBN 978-92-9261-858-2 (print), 978-92-9261-859-9 (electronic)
Publication Stock No. TCS190515-2
DOI: http://dx.doi.org/10.22617/TCS190515-2

The views expressed in this publication are those of the authors and do not necessarily reflect the views and policies of the Asian Development Bank (ADB) or its Board of Governors or the governments they represent.

ADB does not guarantee the accuracy of the data included in this publication and accepts no responsibility for any consequence of their use. The mention of specific companies or products of manufacturers does not imply that they are endorsed or recommended by ADB in preference to others of a similar nature that are not mentioned.

By making any designation of or reference to a particular territory or geographic area, or by using the term "country" in this document, ADB does not intend to make any judgments as to the legal or other status of any territory or area.

Please contact pubsmarketing@adb.org if you have questions or comments with respect to content, or if you wish to obtain copyright permission for your intended use that does not fall within these terms, or for permission to use the ADB logo.

Corrigenda to ADB publications may be found at http://www.adb.org/publications/corrigenda.

Notes:
In this publication, "$" refers to United States dollars.
All photos are from ADB.
This publication was authored by Ms. Sarin Abado (ADB Energy Specialist) under the guidance of Mr. Ashok Bhargava, Director for Energy at ADB's Central and West Asia Department.

Cover design by Jan Carlo Dela Cruz.

CONTENTS

Tables and Figures

TABLES

FIGURES

Gaining energy access.
A light bulb is replaced in a home in Kabul, Afghanistan.

EXECUTIVE SUMMARY

A New Era for Energy in the CAREC Region

The Energy Strategy 2030 for the Central Asia Regional Economic Cooperation (CAREC) program provides a new long-term strategic framework for the energy sector of the CAREC region.[1] It is inspired by the vision of achieving a reliable, sustainable, resilient, and reformed energy market by 2030. Guided by the overarching principle Common Borders. Common Solutions. Common Energy Future., CAREC members are committed to creating a vibrant energy future—a future in which electricity supply is reliable and affordable, energy markets flourish, and cleaner sources have become part of the energy mix.

While the region's energy sector has done well in powering the countries during a challenging period of rapid growth, new dynamics on the global energy scene are rapidly changing the context in which CAREC countries will operate over the next decade. The shifting global dynamics that will influence the CAREC region's energy markets include the rise of renewable energy; the importance of private sector financing in meeting investment needs; the necessity of addressing climate change; and improved political relations, stimulating greater regional cooperation. The CAREC region must be prepared, equipped, and ready to stay ahead of the curve under these new conditions. Hence, a new CAREC Energy Strategy is timely and shall offer fresh approaches and smart solutions enabling the CAREC region to continue to prosper.

The ambitions formulated for 2030 build strongly on past achievements. Today, CAREC member countries are more regionally connected and integrated than ever before. Clean technology deployment is firmly on their agenda, and institutions in the region have greater technical capacity. These achievements provide a solid basis for a bolder vision and a new strategy. With closer regional cooperation, CAREC member countries are well positioned to anticipate and take advantage of the changes occurring within the global energy landscape.

Key Strategic Priorities 2020–2030

Pillar 1: Better Energy Security through Regional Interconnections

Being part of a larger electricity network and gas pipeline system allows energy to be traded at competitive prices, enables diversification of energy sources, and provides more reliable service to consumers. While some CAREC countries are rich in fossil and hydro resources, others lack sufficient domestic resources to adequately cover their energy demand, and seasonal variability among countries is also particularly pronounced.

---

[1] The Central Asia Regional Economic Cooperation (CAREC) program has a membership of 11 countries, with the original eight members being Afghanistan, Azerbaijan, the People's Republic of China (Xinjiang Uygur Autonomous Region joined in 1997; Inner Mongolia Autonomous Region in 2008), Kazakhstan, the Kyrgyz Republic, Mongolia, Tajikistan, and Uzbekistan. Pakistan and Turkmenistan joined in 2010; Georgia in 2016.

The uneven distribution of energy resources and their complementarities provides a strong imperative for regional collaboration. Strengthening cross-border links will allow energy to flow smoothly between countries, reinforcing energy security and economic gains for all.

For energy networks to become more interconnected, the CAREC Energy Strategy 2030 proposes a new regional mechanism for identifying projects of common regional interest through a platform that brings together transmission system operators (TSOs) from all over the region under one umbrella. This new regional TSO platform—the Central Asia Transmission Cooperation Association (CATCA)—shall be set up with a view to allowing network operators to discuss and produce longer-term regional network development plans. The aim of this initiative is to elevate grid expansion planning from a purely national level to a regional level, enhancing information sharing and energy security in the region.

This mechanism introduces a new regional governance system jointly owned by network operators of the CAREC region for the design, development, and operation of the regional grid. Over time, CATCA can also serve as a suitable platform for TSOs to develop region-wide harmonized rules for system operation. These measures will make energy more secure, accessible, and reliable for consumers in the CAREC region.

Pillar 2: Scaled-Up Investments through Market-Oriented Reforms

Energy markets that stimulate competition, attract investment, and generate efficiency across the value chain deliver well-functioning energy services to consumers. In a number of CAREC countries, the energy sector is now undergoing important structural adjustments, moving from a purely state-owned and vertically integrated system to unbundled and liberalized market structures with larger private sector participation. This reform momentum is likely to pick up pace in the next 10 years to achieve a decisive transformation into a modern energy market.

The CAREC Energy Strategy 2030 recognizes that structural reforms are significant undertakings. They require new market designs, governance structures, and overarching regulatory and institutional frameworks. The new strategy is aimed at developing a comprehensive support package to help the CAREC countries in this important endeavor.

The core vision for the next 10 years is to restore financial health in the energy sector in the region by unbundling state-owned companies and introducing modern management practices and cost-reflective tariffs. Finding the right tariff is a true balancing act between competing benefits, such as providing high-quality service at low cost, protecting marginal consumers, and expanding energy access to all consumers. This trade-off becomes even more severe when complementary measures—such as the phaseout of fossil-fuel subsidies, an indispensable tool for leveling the playing field for diverse power sources and achieving financial stability in the energy sector—are introduced.

In tackling the typical dilemmas and trade-offs faced by policy makers in the reform process, capacity building and knowledge sharing are key. The CAREC program will therefore make a particular effort to equip policy makers with practical and tailor-made guidance to allow them to take informed decisions in the complex process of energy sector reform. Emphasis will be placed on the development of measures to protect vulnerable consumers and build social safety nets to ensure a sustainable and socially acceptable reform process. A virtual CAREC Energy Reform Atlas will be a go-to place for concerned stakeholders, providing them with access to practical handbooks and other knowledge materials to answer their questions about how to prepare, implement, and advance energy sector reforms.

Pillar 3: Enhancing Sustainability by Greening the Regional Energy System

The reality of climate change is driving governments and businesses around the world to take urgent action. The region, although a small contributor to global energy-related emissions, is no exception. Various studies have established the region's high vulnerability to climate change. Regional energy efficiency continues to be low because of old and aging infrastructure, low energy pricing, and lack of policy and regulatory support. More efficient energy systems are highly desirable for economic competitiveness, low-carbon intensity, and reliable and affordable energy services to consumers. Rapid deployment of cost-competitive renewable energy and acceleration of energy efficiency are key tools in responding effectively to climate change and for the greening of the regional energy system to enhance its long-term sustainability.

The CAREC Energy Strategy 2030 proposes to keep a clear focus on energy efficiency in the next decade, given that the region holds some of the world's most energy-intensive economies. The CAREC program's role over the coming decade will be to support its members in identifying suitable energy efficiency measures and allow them to become regional champions. A regional Energy Efficiency Scorecard that allows countries to benchmark their progress against international efficiency standards should serve as a central tool for improving energy efficiency in the region besides raising awareness among consumers of how they can save energy.

In the next 10 years, the CAREC program will also provide extensive support to its members in deploying renewable energy, such as solar, wind, and small hydropower, to increase the share of renewable energy in the regional energy mix. This support will include strategies for mitigating intermittency in renewable energy resources and building capacity for suitable incentive schemes to attract more renewable energy generation, preferably through private investments.

To enable the transition toward a green and clean energy sector, the CAREC Energy Strategy 2030 proposes the establishment of a new regional financing vehicle that will allow the CAREC community to mobilize finance for clean energy projects from international and domestic and public and private sources. The CAREC Green Energy Alliance shall be established as a forum for CAREC members to identify and attract these sources of funding. The aim of the alliance should be to create a shared regional fund accessible to all members seeking cofinancing for investments in energy efficiency, renewable energy, and other climate mitigation projects.

Crosscutting Themes

To support the three strategic pillars, the CAREC energy program will also prioritize three crosscutting themes.

Theme 1: Building Knowledge and Forming Partnerships

High-quality data and knowledge sharing are central to operations in the energy sector. Therefore, a number of new knowledge products and partnerships with global centers of excellence shall be established to support the successful implementation of the strategy under its three main pillars.

One major initiative included in the CAREC Energy Strategy 2030 is the launch of a new regional data publication intended to enhance transparency in the state of energy markets in the region. This new regional flagship publication—the CAREC Energy Outlook—will be developed and regularly updated to provide critical insights into regional market trends and contain substantial analyses of what these trends mean for investors, operators, and other stakeholders. The publication will thus be a unique new source of information for investment decision making and significantly support efforts to attract more investments. In delivering new knowledge publications such as the CAREC Energy Outlook, the CAREC program will consider partnerships with established international knowledge leaders and specialized institutions to deliver highly relevant knowledge products.

The CAREC Energy Strategy 2030 also emphasizes the importance of people's networks in fostering regional cooperation, knowledge sharing, and innovation. In this context, new emphasis is placed on young energy professionals in the region to establish a youth platform that gives future energy leaders a louder voice and allows them to participate in shaping the region's energy future.

Theme 2: Attracting Private Sector Investments Across the Energy Value Chain

The energy infrastructure investment needs of the CAREC region (excluding the People's Republic of China) in 2020–2030 are estimated at $400 billion. Currently, the investment levels are about a quarter of the expected level needed and about two-thirds of investments are public sector investments. The public sector can therefore meet only a fraction of the expected investment needs. Success in achieving the desired outcome from the implementation of the CAREC Energy Strategy 2030 hinges largely on creating enabling conditions for crowding in private and commercial capital, to relieve the growing pressure on government budgets.

A regional investment strategy shall be established with suitable partner organizations to guide policy makers in creating effective enabling conditions for private investments. The strategy shall provide recommendations and suggest key elements for a robust and harmonized regulatory framework across neighboring countries. This framework should include facilitation mechanisms for public–private partnerships and the possible establishment of a Central Asia Business Advisory Council to improve the climate for business and strengthen confidence-building measures for investors.

The CAREC program's Energy Investment Forum shall remain the region's flagship platform for actively promoting business–to–business (B2B) exchanges, encouraging start-ups, and drawing in entrepreneurs that can bring new and innovative business ideas into the energy sector.

Theme 3: Empowering Women in Energy

Globally, the energy sector remains one of the most gender-imbalanced, with women making up just 20% of the workforce. The region mirrors this global trend. To make women key drivers of innovative and inclusive solutions in the region, and help accelerate the transition to a clean energy future, concrete actions are needed to improve women's participation in the energy industry and close the gender gap. There are long-term and proven advantages to be derived from having an inclusive and diverse workforce. The CAREC program has a role to act as a champion of women empowerment and lead by example.

To increase women's visibility in the energy sector, the CAREC Energy Strategy 2030 foresees the establishment of a dedicated Women in Energy Program to make women more employable and improve their career opportunities in the sector, besides setting up new regional networking and support facilities for women at all levels of society. These measures should ensure the achievement of the overall vision of gender equity in the energy sector by 2030.

CAREC 2030 Strategy Pillars and Crosscutting Themes

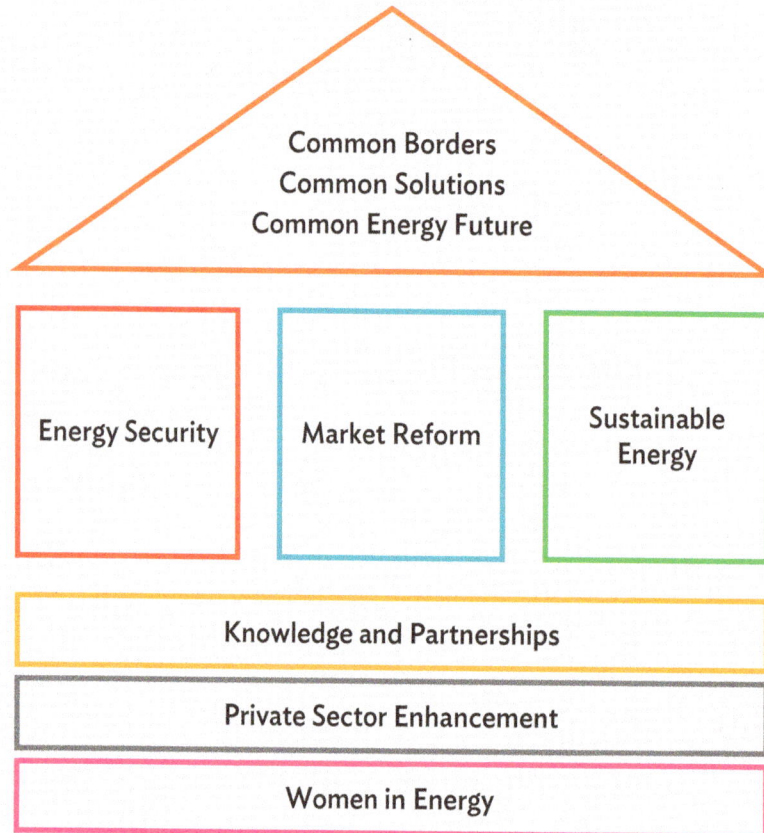

Common Borders
Common Solutions
Common Energy Future

Energy Security

Market Reform

Sustainable Energy

Knowledge and Partnerships

Private Sector Enhancement

Women in Energy

CAREC = Central Asia Regional Economic Cooperation.
Source: ADB.

Delivering Better Results Together

A New Team Spirit for the CAREC Energy Program

All countries concerned must collaborate to implement the CAREC Energy Strategy 2030 and work closely with international financial institutions to achieve the ambitions formulated herein. The regional energy sector endeavors must evolve further into a platform that brings together energy experts, companies, and governments to formulate plans and actions for dealing with emerging challenges. The CAREC program must strive to build a dynamic and high-performing network as the need for knowledge intensifies.

A new team work approach based on working groups and task forces for each of the three strategic priorities and each crosscutting theme shall allow workload sharing and more focused discussions on priority areas by drawing together experts from within the CAREC region. Smart collaboration platforms shall help keep members interconnected and allow them to work together at any given time. This structure shall give CAREC members greater ownership, responsibility, and control of the CAREC Energy 2030 work plan and its deliverables.

Work Plan 2020–2030

The CAREC Energy Strategy 2030 contains a clear work plan of actions to be delivered for each of the three strategic pillars and crosscutting themes. It sets out specific performance indicators for tracking progress. The work plan also proposes a new working group structure for achieving the deliverables, and determines which groups shall be responsible for which actions. In addition to the work plan, the CAREC Energy Strategy 2030 provides a robust results framework, including driving principles, outcomes, and interventions, for attaining the overarching goals of the strategic pillars and crosscutting themes.

The vision for 2030 is wide-ranging in scope but achievable. It seeks a smarter, efficient, green, sustainable, and resilient energy system for the region. A reliable energy system is the backbone of any modern economy and essential for the well-being of people. With common borders, common solutions, and a common energy future, a new and promising era for energy markets in the CAREC region is on the horizon.

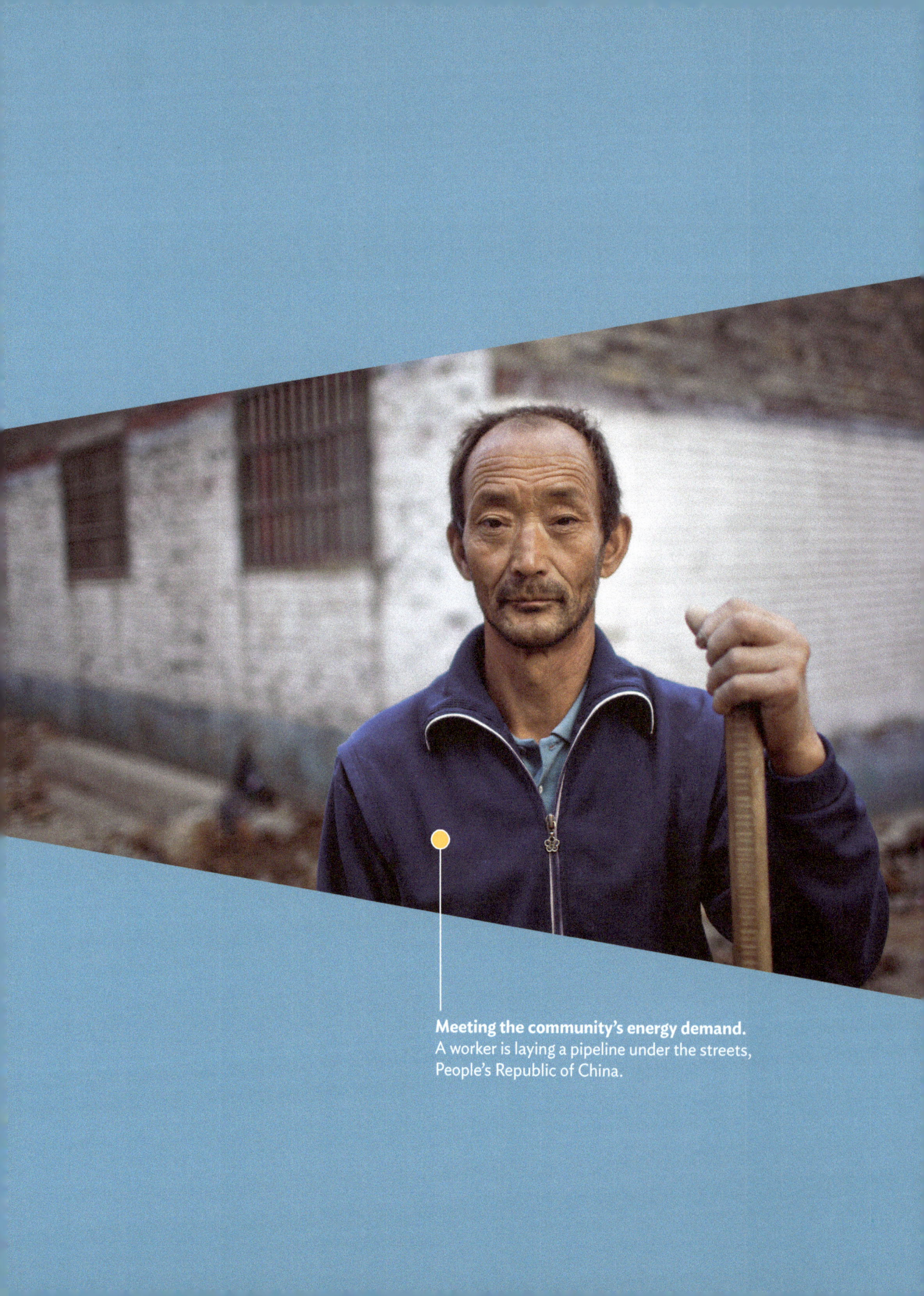

Meeting the community's energy demand.
A worker is laying a pipeline under the streets,
People's Republic of China.

I | RATIONALE AND DRIVING PRINCIPLES FOR A NEW CAREC ENERGY STRATEGY 2030

The Need for a New CAREC Energy Strategy 2030

The energy system of a country is one of its most complex, expensive, and sophisticated sets of assets. It provides electricity and gas to millions of homes, offices, industrial complexes, and many other essential services. A reliable energy system is the backbone of any modern economy and essential for the well-being of people. "Keeping the lights on" under all conditions is therefore a shared challenge for the entire Central Asia Regional Economic Cooperation (CAREC) region[1] and remains a top priority up to 2030.

The energy sector has done well to power the countries during a challenging period of rapid growth, but new dynamics on the global energy scene are increasingly changing the context in which CAREC countries will operate over the next decade. The future global energy system will look different from the past. Therefore, a new CAREC Energy Strategy, offering smart solutions and new approaches, is timely. It will allow the CAREC region to stay ahead of the curve.

The ambitions formulated for 2030 shall build strongly on the achievements of recent years. Today, CAREC member countries are more regionally interconnected and integrated than ever before and are benefiting from increases in cross-border energy trade. Clean technology deployment is firmly on the agenda and there is greater technical capacity of governments and institutions. These achievements provide a strong foundation for a new strategy.

Overall, the energy sector in the CAREC region is at an important crossroad. Global trends in the climate, economic, geopolitical, and financial fronts are reaching the local and regional levels, creating both challenges and opportunities for the region. The key factors that will influence energy markets in the CAREC region in the next decade are the following (Figure 1):

- **Renewable energy is becoming a real alternative.** Solar power and large-scale wind generators have undergone rapid deployment and begun to compete on costs with fossil-fuel–based technologies. The total capacity of renewable power installed globally has outstripped fossil-fueled capacity for the past 4 years, with higher investment in developing countries than developed countries.[2] Countries with high levels of renewable energy are successfully addressing the challenge of integrating variable renewable energy sources into electricity grids.

---

[1] The Central Asia Regional Economic Cooperation (CAREC) program has a membership of 11 countries, with the original eight members being Afghanistan, Azerbaijan, the People's Republic of China (Xinjiang Uygur Autonomous Region joined in 1997; Inner Mongolia Autonomous Region in 2008), Kazakhstan, the Kyrgyz Republic, Mongolia, Tajikistan, and Uzbekistan. Pakistan and Turkmenistan joined in 2010; Georgia in 2016.

[2] Renewables 2018. Global Status Report. REN21. http://www.ren21.net/gsr-2018/; and Renewables 2017. Global Status Report. REN21. https://www.ren21.net/wp-content/uploads/2019/05/GSR2017_Full-Report_English.pdf.

Figure 1: Driving Principles for a New CAREC Energy Strategy 2030

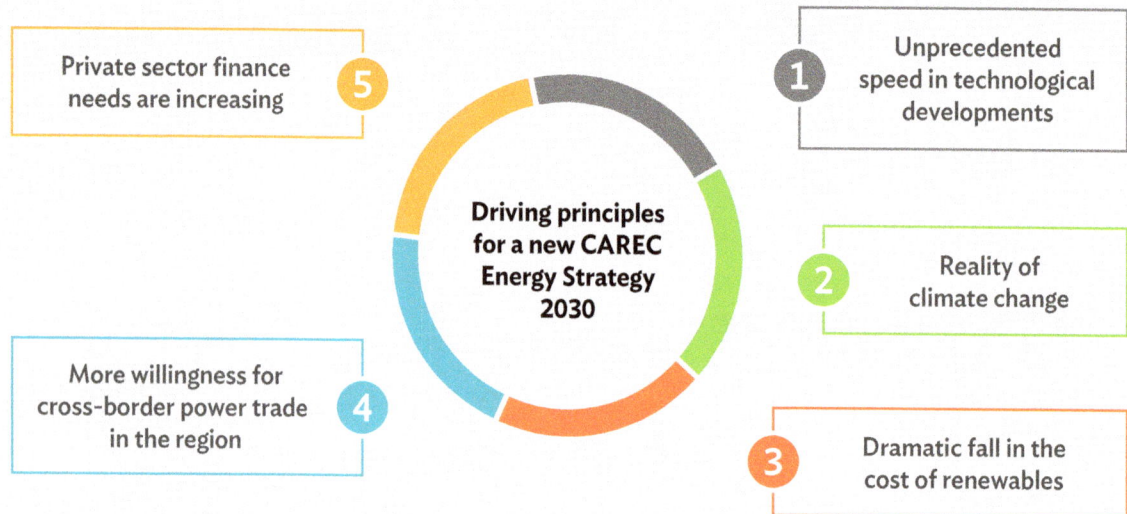

Driving principles for a new CAREC Energy Strategy 2030

5 Private sector finance needs are increasing

1 Unprecedented speed in technological developments

2 Reality of climate change

4 More willingness for cross-border power trade in the region

3 Dramatic fall in the cost of renewables

CAREC = Central Asia Regional Economic Cooperation.
Source: ADB.

- **Climate change is impacting policy decisions.** The energy sector is a major contributor to regional carbon dioxide (CO_2) emissions. The CAREC region is highly vulnerable to the effects of climate change, such as rising temperatures, water shortages, and extreme weather events. These pose a serious threat to physical infrastructure in the region. Mitigating climate change and building resilience to its impacts must be important considerations in energy investment decisions throughout the region. Globally, over 140 countries have set national targets for renewable energy in their power sector.[3] The greening of the power sector through the rapid deployment of renewable energy and acceleration of energy efficiency is key in formulating a climate change–responsive approach to energy sector planning in the CAREC region.

- **The global economic downturn is depleting state budgets.** Energy and commodity exporters are particularly affected by the ongoing downward pressure on prices from dampened demand and the worldwide oversupply of energy. Regional economies, especially fossil-fuel–rich members such as Azerbaijan, Kazakhstan, Turkmenistan, and Uzbekistan, which rely predominantly on hydrocarbon exports to fill their state budgets, are at severe risk. In addition, most CAREC countries offer generous energy subsidies to their consumers, which further drain government budgets and are neither financially sustainable nor in the long-term interest of consumers.

---

[3]   Renewables 2019. Global Status Report. REN21. https://www.ren21.net/wp-content/uploads/2019/05/gsr_2019_full_report_en.pdf.

- **Necessity of unlocking private capital to meet large investment needs.** Energy sector investment needs in the CAREC region (excluding the People's Republic of China) up to 2030 are estimated at $400 billion. Governments provide about two-thirds of investments currently but will not be able to meet their projected needs up to 2030 alone. Thus, larger private and commercial capital is essential to fill the financing gap. Attracting private capital requires market-oriented reforms and better sector governance. Given the recent reform momentum in the region, there is renewed hope for larger private investments.

- **Political relations among CAREC countries are improving.** Positive political changes are occurring in the region. These have created fresh opportunities for the member countries to cooperate on a myriad of long-standing problems. There are also encouraging signals from the Caspian region that the decades-old Caspian Sea conflict is headed for resolution, opening up possibilities for greater regional cooperation.

In light of these developments, the CAREC region must be prepared, equipped, and ready to act to advance its energy markets under these new conditions. A new CAREC Energy Strategy 2030 is therefore timely and presents an opportunity to create a clear and robust vision for the future.

There is scope for development in the following key strategic areas up to 2030:

- greater regional interconnection, to improve energy security;

- more competitive, market-oriented approaches, to deliver greater benefits to consumers;

- enhanced sustainability, through the greening of the energy system;

- new knowledge and partnerships, to deepen and foster long-term regional relationships;

- greater participation from the private sector; and

- equal opportunity for women to build a diverse and vibrant sector.

The CAREC Energy Strategy 2030 is a fit-for-purpose and future-proof energy strategy for the Central Asian region that will position the region to meet the challenges of the changing regional and global energy landscape and emergent risks and opportunities.

Regional energy connectivity. Cross-border power lines between Uzbekistan and Afghanistan.

Building a New Era for Energy in the CAREC Region

Driven by the Central Asia Regional Economic Cooperation (CAREC) spirit of Good Neighbors. Good Partners. Good Prospects., member countries are committed to the development of the region's energy sector to ensure its long-term economic competitiveness. Guided by the overarching principle of Common Borders. Common Solutions. Common Energy Future., members are inspired to work together for a common vision of a reliable, sustainable, resilient, and reformed energy market by 2030 (Figure 2).

Figure 2: CAREC Region's Energy Vision 2030

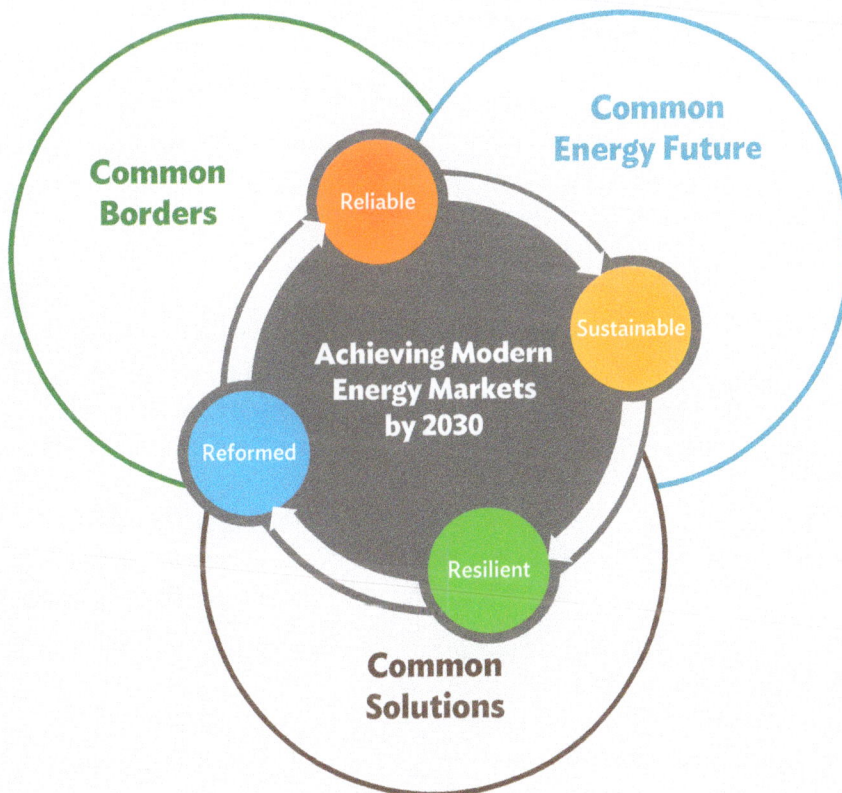

CAREC = Central Asia Regional Economic Cooperation.
Source: ADB.

To ensure that energy is available, reliable, and financially and environmentally sustainable, CAREC members envisage a vibrant energy future:

- **The Future Leaves No Customer Behind**

  There is a commitment to share and expand the regional cross-border energy grid for maximum energy security. Gas and electricity should flow where they are needed and deliver affordable and sufficient supplies to all corners of the region so that no customer is left behind.

- **The Future Is More Market-Oriented**

  There is a commitment to transform the energy sector from a government-owned and government-operated model, with vertical integration and subsidies, to a more market-oriented one. There is a pledge to build stronger institutions and regional market governance to support the process of market liberalization.

- **The Future Is a Green Energy Mix**

  To address the challenges posed by climate change and ensure economic competitiveness, there is renewed attention to addressing energy efficiency and seeking a more diversified energy mix with a higher share of clean and renewable energy. Regional integration is also sought, to leverage more cost-competitive energy storage options.

- **The Future Is Inclusive and Equitable**

  The region is striving to develop the next generation of energy professionals, who are resourceful and skilled, and to increase the engagement and visibility of women in the energy sector. The sharing of knowledge and expertise across generations, borders, and disciplines provides a foundation for long-term cooperation in a closely interconnected region.

The vision for 2030 is wide-ranging in scope, but achievable. It recognizes that the region's energy and economic future are closely interlinked. Therefore, the vision for 2030 can be successfully achieved only if relevant energy and economic transitions are actively pursued over the next decade. The CAREC program should provide a vehicle and platform for planning these transitions:

- **Transition to new fuels and technologies.** The replacement of high-emission generation (such as coal production) with lower-emission alternatives, including hydro, gas, and renewable energy technologies, is where the energy industry is heading. Renewable energy, such as solar and wind, has increasingly become more competitive as costs have drastically reduced. Hence, these new technologies must be actively considered as part of any new generation capacity development plan. Intensive research and development underway internationally on energy storage solutions will improve their deployment outlook even further in the CAREC region. To aid the transition to a new energy mix, CAREC member countries should work toward removing barriers to the uptake of cleaner fuels, through such means as the phaseout of fossil-fuel subsidies.

- **Transitions in energy market designs and institutional structures.** Fuel and technology transitions have given rise to an energy sector that is rapidly becoming more complex. The ability of energy market institutions to ensure that policy and regulatory settings remain relevant will be critical, as will their ability to work collaboratively to manage the energy transition and develop new knowledge and internal capability. To deliver on the 2030 vision of a reliable, sustainable, resilient, and reformed energy market, there is an urgent need for institutional capacity development and a comprehensive energy policy reform agenda.

- **Industrial transitions.** Energy exports in the past decades depended on hydrocarbon-rich countries delivering their resources to importing countries as energy commodities. The trading of electricity was, however, constrained by the lack of integrated regional markets operating on market principles. With the anticipated increase in the uptake of renewable energy, and a more sophisticated and integrated market structure and electricity pricing regime, new electricity export opportunities will emerge for countries rich in renewable energy. There is also an anticipated shift in economic activity occurring among suppliers of new technologies: solar, wind, storage, electronic controllers, and electric vehicles. Given the nascent status of these technologies in the region, there has been an economic uptick in collective assessments of regional industry development opportunities, considering existing industry strengths, capabilities, and resources.

2020–2030: Decade of Transformation

The implementation of a shared vision by CAREC member countries will help in setting up a modern energy future in which end-use consumers are more informed about efficient and clean energy and more incentivized by economic signals. A transition to a new energy future will bring long-term benefits to all energy consumers in terms of availability, quality, and security of energy supply. At the same time, the transition will deliver far-reaching benefits to regional economies in terms of new investment, jobs, skills, and knowledge development (Figure 3).

A future-proof energy sector for the CAREC region is on the horizon. With regional collaboration and cooperation, CAREC member countries will be well positioned to anticipate and take advantage of the broader changes occurring within the global and regional energy landscape.

Figure 3: Energy Sector Transformation in the CAREC Region 2020–2030

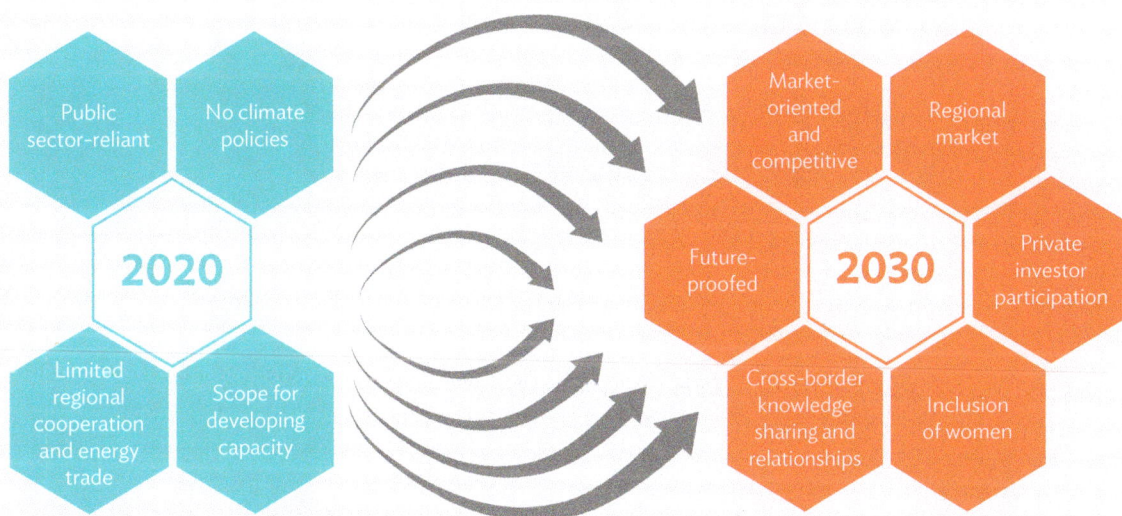

CAREC = Central Asia Regional Economic Cooperation.
Source: ADB.

Building energy infrastructure. Local workers help convoy to pass under high-voltage cable, Uzbekistan.

KEY STRATEGIC PRIORITIES 2020–2030

PILLAR 1

Better Energy Security through Regional Interconnections

Cross-border energy interlinkages between Central Asia Regional Economic Cooperation (CAREC) countries have significantly improved over the past decade. The spillover benefits of such connections are evident. Large-scale energy infrastructure achieves economies of scale, instills a collaborative culture, and generates a strong drive toward common energy security through long-term regional relationships.

Mutual Benefits from Cross-Border Energy Connectivity

Being part of a larger grid network and pipeline system has several advantages. It allows energy to be traded at competitive prices, enables diversification of energy sources and routes and provides more reliable services to consumers as support from neighbors can quickly compensate for any system imbalances. A densely meshed grid is also key in successfully bringing intermittent renewable energy sources online. Coupled with appropriate regulatory reforms, interconnections serve as catalyst to accelerate market integration and connections to global value chains.

While most CAREC countries are rich in fossil and hydro resources, other countries, particularly in South Asia, face challenges in meeting energy demand from domestic resources. The uneven distribution of energy resources in the region and their seasonal complementarities provide a strong imperative for regional collaboration and partnerships. Hence, creating smart links throughout the region to allow energy to flow from countries with excess supply to those in high demand will reinforce energy security and economic gains for all.

Regional interconnections are also vital to building the confidence essential for investments and easing political constraints. The linking of natural gas and electricity industries across borders can indeed justify mobilizing investments by expanding market size. Increased competition and lower costs lead to a win–win situation for all countries involved, creating a more conducive environment for broader economic integration and trade.

Launching Mega Projects of Common Regional Interest

Recent steps taken to expand cross-border linkages have been encouraging. These augur well not only for the reintegration of the Central Asia Power System (CAPS) but also for its likely expansion to new markets, such as Afghanistan and Pakistan. A reinvigorated CAPS can allow more market-based approaches to cross-border electricity trade and better integrate a larger anticipated share of renewable energy into the regional energy mix.

A number of other cross-border energy infrastructure projects, mostly linking Central Asian energy producers with South Asian consumers, are now being implemented in the CAREC region. These projects are at various stages of implementation and include the following flagship electricity and gas interconnection projects: TUTAP, TAP, CASA-1000, and TAPI.[4]

The CAREC program shall step up its efforts to make these operational as early as possible and track progress on these projects. Further regional connections to achieve supply–demand balance at competitive prices will also allow the region to build links with other attractive energy markets outside the region, e.g., with East Asia and Europe, in the long run.

As a new vehicle for promoting regional interconnections, projects of common regional interest shall be identified and planned together by the CAREC members, ideally through the development of longer-term regional network development plans. This will not only send clear signals to investors but also give transmission system operators the opportunity to develop a common long-term vision for a shared regional energy grid that operates on agreed principles.

Working Together in Regional Institutions

There are strong reasons why CAREC countries should work together not only in building and operating physical interconnections but also in creating appropriate regional governance structures for deciding on issues related to grid expansion, modernization of energy infrastructure, and other operational issues such as data sharing.

CAREC members shall therefore strive to develop a concept for a regional body—the Central Asia Transmission Cooperation Association (CATCA)—bringing together all transmission system operators from the region to plan the development of the regional grid (as opposed to their national grids alone) and to develop region-wide rules and standards for operating the grid. This body shall ideally have the following functions:

- elaborating a long-term regional network development plan including projects of common regional interest to receive priority support;

- providing regular information on electricity and gas supply and demand for the market; and

- developing harmonized region-wide rules for system operation.

This body shall be composed of transmission system operators from each country and provide appropriate structures for building consensus in the decision-making process (Figure 4). It shall also be established with a view to fostering energy trading on a longer-term horizon.

As interconnections increase, CAREC countries will become more interdependent and collectively responsible for preventing disruptions. Any local incident may have repercussions in neighboring countries and should therefore also be resolved collectively and in a spirit of solidarity.

---

4   TUTAP stands for Turkmenistan–Uzbekistan–Tajikistan–Afghanistan–Pakistan; TAP for Turkmenistan–Afghanistan–Pakistan; CASA for Central Asia–South Asia; and TAPI for Turkmenistan–Afghanistan–Pakistan–India.

Figure 4: New Governance Structure for Regional Network Development Planning

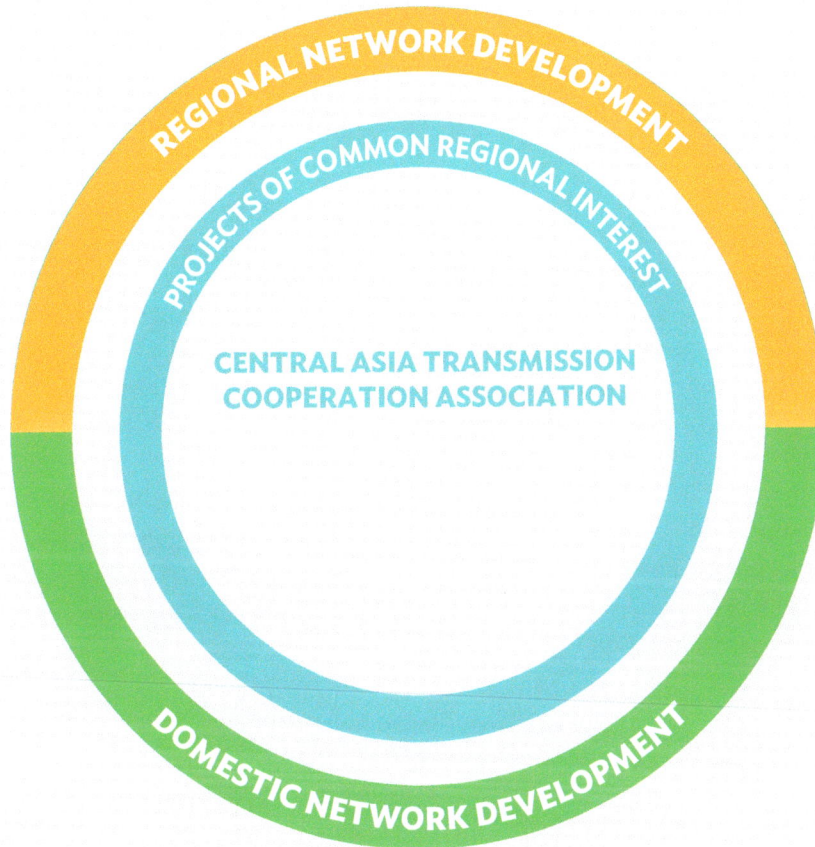

REGIONAL NETWORK DEVELOPMENT

PROJECTS OF COMMON REGIONAL INTEREST

**CENTRAL ASIA TRANSMISSION
COOPERATION ASSOCIATION**

DOMESTIC NETWORK DEVELOPMENT

Source: ADB.

Achieving Universal Access to Energy Across the Region

Achieving Sustainable Development Goal 7—access to affordable, reliable, sustainable, and modern energy for all—is an important regional goal. At the outset, it may appear to be a national, rather than a regional, endeavor. However, the electricity and gas needs of populations in border towns are, in many instances, better and more cost-effectively served by a neighboring country than through extensions of the national grid. While the majority of consumers in the region are connected to the grid, a number of mountainous and rural areas remain difficult or expensive to reach through grid infrastructure. Hence, off-grid systems, preferably powered by renewable energy, shall be evaluated for rollout in communities in need. To this end, the experience and lessons learned from pilot projects in and outside the region shall be shared widely. By 2030, consumers in all corners of the CAREC region, whether connected to the grid or not, should enjoy access to modern electrification.

PILLAR 2
Scaled-Up Investments through Market-Oriented Reforms

Competitive energy markets bring down costs, stimulate innovation and generate efficiency to the benefit of consumers. Energy market liberalization is still at an initial stage in the majority of CAREC member countries and requires relevant reforms over a continuous period of time.

The Power of Reforms

Despite recent progress, traditional government control of electricity and gas companies, as well as the absence of competition, has led to distorted energy prices, inefficient network operation, and deteriorating infrastructure in many parts of the CAREC region. The energy sector's deep-seated financial difficulties have undermined its ability to invest in existing and new productive assets. The lack of financial resources in the energy sector are mainly caused by tariffs requiring subsidies and substantial fiscal support in many parts of the region, leading to inadequate generation of funds for the sustainable maintenance of existing assets.

Sector reforms have the power to break this cycle and establish more market-oriented structures with state-owned enterprises operating on the basis of principles similar to those that drive private companies. This approach will lay the groundwork for competition and market mechanisms to evolve, and thus respond to the expectations of end users for high-quality service and affordable prices (Figure 5).

For energy sector reforms to work, they must be enshrined in law with appropriate enforcement mechanisms to ensure that actors comply or else face consequences. This includes clarifying the roles and responsibilities of government versus utilities and other actors in the market. In this important process of adapting existing legislation and creating new laws, the CAREC program shall provide a platform for its members to discuss successful international examples of sector legislation and law enforcement methods.

Paving the Way for Structural Reforms

Vertically integrated value chains still prevail in a large number of CAREC countries. Separating the individual activities into distinct entities for generation and network-related activities will enhance the performance of the sector, as each entity will have to sustain itself financially without backup and cross-subsidies from the vertically integrated structure.

The benefit of unbundling consists in the possibility of opening up some parts of the value chain to competition, especially energy generation. It also forces the remaining parts of the value chain, i.e., network-related activities that remain state-owned, to ensure sufficient income on their own to build, strengthen, and maintain the grid. The condition of energy networks, both transmission and distribution, has deteriorated in the past decades due to lack of financial means and effective incentives. Unbundling will give network operators and distribution companies an incentive to deliver an optimal level of service quality at the least cost.

Unbundling reforms shall be initiated as a first step in creating the fundamental structure for energy markets to move toward liberalization. The greater the degree of separation, the higher the incentives for the network operator to perform high service quality and the easier for new generation companies to enter the market.

Figure 5: Benefits from Market Liberalization

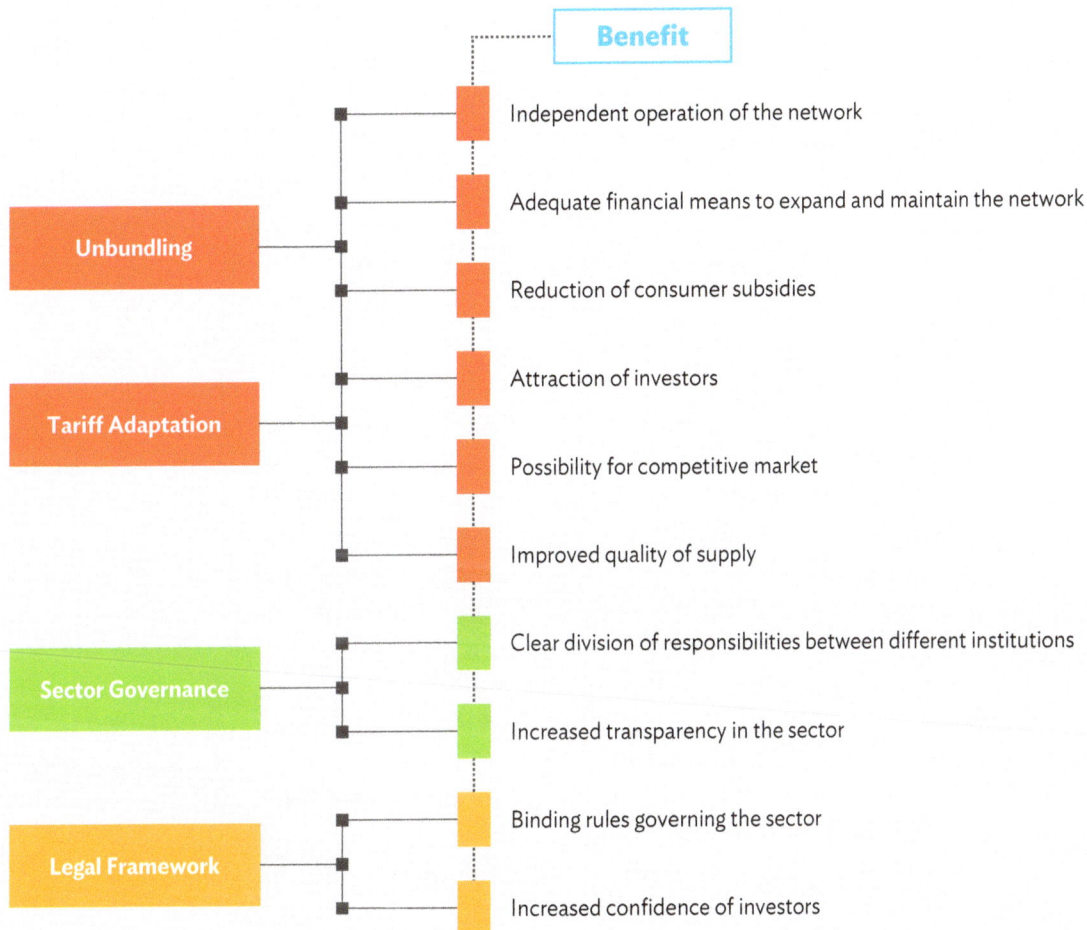

Benefit

Independent operation of the network

Adequate financial means to expand and maintain the network

Reduction of consumer subsidies

Unbundling

Attraction of investors

Tariff Adaptation

Possibility for competitive market

Improved quality of supply

Clear division of responsibilities between different institutions

Sector Governance

Increased transparency in the sector

Binding rules governing the sector

Legal Framework

Increased confidence of investors

Source: ADB.

For the CAREC region to embark on unbundling and market liberalization reforms, capacity building and knowledge sharing are essential. Reforms will succeed only if carried out based on informed decisions and models that fit the individual national context. Therefore, the CAREC program shall establish a virtual Energy Reform Atlas as a go-to place for information about unbundling models and other aspects of market liberalization, to support CAREC members in the reform process. The atlas shall create a solid knowledge base including a web-based directory for study materials that are accessible at any given time to policy makers in the region.

Modern Tariffs for Enhanced Financial Sustainability

Finding the right tariff is a balancing act, as users want high quality of supply at a low cost. While current end-user tariffs in many CAREC countries are low, quality of service lags behind because the low tariffs do not reflect the actual cost incurred in running and maintaining the network, let alone crucial investments needed to perform quality service over a longer stretch of time.

To determine the level of tariffs that will allow performing at a high standard, a careful and sophisticated cost auditing procedure to determine adequate cost levels is needed. Once the cost base is determined, a "fair" rate of return on capital (cost recovery principle) is established and added to arrive at the optimal price. This rate-of-return, or cost-plus tariff principle is a classical approach used in many countries, especially at the start of unbundling and market liberalization reforms.

Tariff setting is a complex domain that requires adequate skills and awareness of different principles that exist, including best practice examples. The CAREC program shall therefore establish a practical guide for its members in the form of a handbook that contains advice on tariff-setting principles and methods to introduce tariff reform. Customized capacity building workshops shall also be held to address frequently asked questions and develop appropriate models for the target audience.

Gradual Removal of Energy Subsidies

Energy prices, costs, and subsidies are among the critical issues that determine the health of any economy. Today, most consumers in the region receive generous energy subsidies directly or indirectly, regardless of income. The average consumer electricity tariffs are among the lowest in the world. The resulting fiscal support needed is a substantial burden on the economy in many countries in the region which is neither financially sustainable nor in the long-term interest of consumers. The high share of low-income households in the CAREC region's population makes tariff policy reforms a politically difficult task, but experience shows that a variety of measures exist to effectively protect vulnerable consumers.

First, fossil-fuel subsidies should be gradually phased out given that they keep fossil fuels in hydrocarbon-rich economies at an artificially low price in the domestic market compared with international pricing, and in many cases even lower than marginal cost of production. The practice encourages wasteful use of energy, drives up greenhouse gas emissions, inhibits the promotion of energy efficiency, and introduces an uneven playing field for renewable energy. Subsidized carbon-based fuels have delayed the deployment of renewable energy in the region and have locked the region in energy- and carbon-intensive economies.

The removal of fossil-fuel subsidies can go hand in hand with comprehensive tariff reforms. It is a process that requires time, diligent planning, and consideration of social aspects. Therefore, subsidy reform accompanied by other economic and social reforms can ensure that the overall goal of stable and affordable energy is met. The CAREC program will develop a suite of policy notes and knowledge pieces to help and guide members in taking on complex yet critical reforms tailored to individual country contexts.

Protecting Marginal and Vulnerable Consumers

Energy sector reform not only affects the economy and markets but also has a strong impact on consumers. Reforms are often delayed or avoided as they are evaluated narrowly on the basis of tariff impacts alone. While removing subsidies and distortions will invariably lead to higher tariffs, well-designed reforms will also free the government from its fiscal burden and allow it to spend more on social services, such as education and health. Moreover, a targeted policy approach accompanying tariff reforms can make sure that energy does not become unaffordable to large parts of the population, and that marginal and vulnerable consumers are well protected.

The CAREC program shall thus help its members to develop a definition of vulnerable energy consumer and develop options for social protection measures. There are many ways in which vulnerable consumers can be supported and such measures have been successfully introduced in many economies. Appropriate knowledge products and policy briefs, based on international experience and lessons, will be developed to help policy makers understand the various options and trade-offs in making appropriate and informed decisions.

PILLAR 3
Enhancing Sustainability by Greening the Regional Energy System

The reality of climate change is challenging governments and businesses around the world to take urgent action. The region, although a minor contributor to global energy-related emissions, is no exception. Overall, two-thirds of greenhouse gas emissions result from energy production and use alone.[5] Spreading awareness of clean energy and energy saving is of paramount importance in greening the regional energy system for long-term sustainability.

Energy Efficiency as a Fuel of First Choice

The implementation of the Paris Agreement adopted during the United Nations Conference of Parties (COP21) in 2015 has become a global driver for bringing down greenhouse gas emissions, accelerating the deployment of renewable energy, and achieving significant energy savings. Indeed, prioritizing energy efficiency has a number of advantages as it is a cost-effective means to reach a low-carbon economy and a fuel of first choice in improving competitiveness in an interconnected regional market (Figure 6).

Figure 6: Holistic Approach to Achieving a Sustainable Regional Energy System

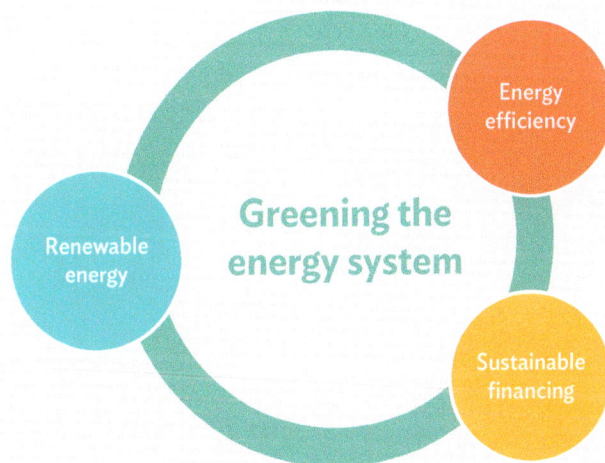

Energy efficiency

Renewable energy

Greening the energy system

Sustainable financing

Source: ADB.

---

[5]   Good Practice in Energy Efficiency. For a Sustainable, Safer and More Competitive Europe. European Union. https://ec.europa.eu/energy/sites/ener/files/documents/good_practice_in_ee_-web.pdf.

Energy intensity levels and trends differ widely across the world, reflecting differences in economic structure, technology maturity, and energy efficiency achievements. The level of energy efficiency in the CAREC region is relatively poor compared with global averages because of dilapidated infrastructure, subsidized energy prices, and weak regulations and policy support. While newly built infrastructure is modern and energy-efficient, the retirement of old infrastructure is hardly a policy priority. Thus, overall energy efficiency remains low.

Energy efficiency has huge untapped potential in the region and is a priority in the nationally determined contributions (NDCs) of most countries. Over the next decade, the region can effectively take the fuel of first choice approach to energy efficiency to drive it up to international standards.

Making energy efficiency a policy priority requires the right mix of regulations, incentives, and energy efficiency considerations in the decision-making matrix for new investments. A right mix of these necessary ingredients can generate pull for energy efficiency investments. Given the prevailing high energy intensity across the CAREC region, energy efficiency improvement is a potent low-cost, high-impact measure for reducing carbon emissions.

In this context, the CAREC program's role over the coming decade should be to support its members in identifying suitable energy efficiency measures and creating suitable business models that will allow investments to flow in this area. To this end, a regional CAREC Energy Efficiency Scorecard shall be developed to allow countries to benchmark their progress against international efficiency standards. With this benchmarking tool, countries can identify gaps and formulate suitable measures for bridging those gaps. Above all, the scorecard should motivate CAREC members to become energy efficiency champions who can share their experience with their peers.

Yet the more fundamental ingredient for achieving long-lasting energy efficiency is creating widespread knowledge and true awareness of it. Consumers and business owners must be empowered with the knowledge of how they can save energy in their daily lives and business operations. Therefore, in the coming decade, the CAREC program will develop recurring public and region-wide energy efficiency campaigns including radio and TV commercials and a CAREC consumer leaflet about energy savings. In addition, a CAREC Energy Efficiency Week shall be held across the region to enhance public awareness of this subject. A reference guide to effectively organizing consumer awareness campaigns shall also be developed to enable governments and other stakeholders to repeat the exercise on their own whenever deemed necessary.

Creating a Cleaner Energy Mix and Economic Diversification

A key solution to the challenge of reducing energy intensity and carbon emissions is to progressively bring down the share of fossil fuels in the generation mix and seek a more diversified mix with a higher share of clean and renewable energy sources (Figure 6). Despite abundant renewable energy potential in the CAREC region, including small hydro, wind, and solar energy, installed solar and wind energy capacity currently amounts to less than 1% of the total capacity. While many countries in the region have announced ambitious targets for renewable energy development, both solar and wind are nascent industries in the region and face many perceived and real risks in this early stage of deployment.

With the costs of solar and wind power generation rapidly decreasing, countries are setting up policy support to attract private capital for the deployment of these resources. International financial institutions that have assisted governments in formulating such policies and in launching renewable energy auctions are working closely with governments to get the countries on the right trajectory of renewable energy deployment. Initial results in this endeavor over the past 2–3 years across the region have been very encouraging. However, integrating a larger share of renewable energy into any country's grid is still perceived as a major risk in the CAREC region.

Within the next 10 years, the CAREC program will provide support to its members to attract more renewable energy generation in the regional energy mix, preferably through private investments. This support will include addressing the technical challenges connected with the intermittent supply of renewable energy as well as providing guidance on various renewable incentive mechanisms, their pros and cons, and financing solutions for these incentive schemes. The aim is to instill in policy makers the confidence to make informed decisions on how to achieve the energy transition effectively in their respective countries. Policy makers will be equipped with a robust guide on legislative and technical requirements for integrating renewable energy into the grid. The CAREC program will ensure that this knowledge transfer is realized through tailor-made capacity building programs and opportunities for the global exchange of best practices, especially those developed in countries like Germany and others that have come a long way in the energy transition.

A smooth transition toward renewables and a low-carbon economy will require factoring in the role of transition fuels, especially gas. Gas has significantly lower carbon emissions on combustion per unit of energy delivered than either coal or oil and may therefore act as a "bridging fuel" to a low-carbon energy future. Given that the creation of a new and more renewable energy mix will be gradual and will take place over a long stretch of time, a focus on gas (which is abundant in the CAREC region) should be included in the regional debate around the energy transition to ensure that the transition is realistic and sustainable. Given the region's acute vulnerability to the impact of climate change, smart adaptive approaches and measures that will enhance the resiliency of existing and new energy assets must be an important consideration.

Securing Innovative Financing for a Smooth Transition

Deploying renewable energy and energy efficiency measures is capital-intensive and thus poses formidable challenges to attracting financing at scale. Given the historic level of predominantly public sector–led investments in the energy sector, unlocking and crowding in private and commercial financing is absolutely crucial for meeting the investment needs of the next 10 years across the region. Innovative approaches, such as guarantees, leasing, equity, and debt financing from diverse sources like commercial, private, and institutional investors, are key.

Leading financial institutions and large wealth funds in the region are highly interested in impact financing, which combines adequate returns on socially and environmentally responsible projects. International financial institutions have a strong commitment to enlarge climate financing. Thus, there is a need to develop a pipeline of investment-ready projects and link them to financing mechanisms from a variety of sources.

To this end, the CAREC program shall establish a joint platform—a green-energy marketplace—to bring together project developers and potential financiers. The CAREC Green Energy Alliance can be such a forum where all CAREC members can identify and attract sources of funding for a pipeline of well-developed and investment-ready green and climate-responsive energy projects. The long-term aim of the alliance should be to create a shared regional fund accessible to all members that will provide them with end-to-end solutions for cofinancing investments in clean energy projects (Figure 6).

Energy for all.
Woman in her home powered
by solar and wind electricity,
Pakistan.

IV | CROSSCUTTING THEMES

To support, complement, and reinforce its three main strategy pillars, Central Asia Regional Economic Cooperation (CAREC) members will also prioritize several crosscutting themes. These will improve the results and outcomes of the main strategy pillars, fill critical gaps, and ensure the successful implementation of the CAREC region's 2030 vision for the energy sector. The envisaged crosscutting themes are (i) building knowledge and forming partnerships; (ii) enhancing the role of the private sector; and (iii) empowering women in energy.

THEME 1
Building Knowledge and Forming Partnerships

Knowledge is central in the transition to a future-oriented energy sector in the region. Therefore, a number of knowledge products and partnerships with knowledge institutions shall be established to support the successful implementation of the three main strategy pillars. The next generation of energy professionals will be encouraged to participate in all knowledge-related activities. Knowledge sharing within the CAREC region is another key element of promoting cooperation in the regional energy sector.

A New CAREC Energy and Investment Outlook

In response to requests from investors, governments, international organizations, and civil society, the CAREC program shall develop an Energy and Investment Outlook at periodic intervals. Based on comparable data and analysis gathered from member countries and other organizations, the document will provide critical insights into trends in energy demand and supply, within a wider economic and demographic context. The outlook shall also contain substantial analysis of what the identified trends mean for investors, operators, and other stakeholders. It shall specifically examine the impact of current and future trends on energy security, environmental protection, and economic development in CAREC countries.

A specific energy investment chapter in the document shall report progress on regional and international investment flows to the energy sector. It should also shed light on new opportunities for, and barriers to, investment. Input for this report shall come from the relevant business communities, among other sources.

These data-driven products will provide enhanced transparency regarding the state of the region's energy markets and support efforts to attract more investments, track the reform process, and achieve top-quality service for consumers.

Developing the CAREC Region's Young Energy Leaders

Strengthening the role, competence, and voice of young professionals is investing in the future. Millennials and post-millennials are growing up in a world of accelerated energy transformation and are more aware than ever of climate change, clean energy solutions, and the possibilities of technological leapfrogging. Yet this generation is still underrepresented at the decision-making level in the energy industry and is hardly involved in designing new business models, the rules of the game, and the new mind-sets that are required to prosper in the future. The idea of playing a leading role in the clean energy transition and being at the forefront of a groundbreaking technology and industry is, however, clearly attractive to many young people.

Millennials are often reluctant to enter the hydrocarbon industries because they are concerned about the long-term prospects of these industries. Given that oil, gas, and hydropower represent the competitive strength of CAREC countries today, there is a risk that unless the industry attracts young talent, its medium- to long-term viability will be further eroded. Listening to the ideas, motivations, and visions of the next generation is therefore crucial in building and sustaining a modern energy sector in the CAREC region.

Bringing together young talent from the region has a huge payoff as it will not only allow young professionals to meet and work in joint projects across the region and give them a stronger voice in the energy sector but also enhance regional cooperation and integration efforts. The CAREC program shall therefore provide a regional youth platform for discussions about energy security, infrastructure, climate change, energy efficiency, and gender equality, as reflected in the main pillars of this strategy.

Chosen by virtue of their potential to become energy-industry leaders, young professionals in the CAREC region shall form a unique community and take part in thought leadership discussions led by prominent executives and officials. To this end, the CAREC program shall launch a Young Energy Leaders Initiative designed for young professionals from the government, private companies, universities, and nongovernment organizations to help talented youth in the CAREC region take leadership roles in the energy world of tomorrow. In particular, the initiative shall provide opportunities for networking, skills development, and practical exposure to various parts of the industry.

Working with New Partners

The comprehensive vision for 2030 opens up new possibilities for partnerships between CAREC members and professional institutions. In fact, it is neither feasible nor desirable for the CAREC program to undertake all the initiatives proposed in this strategy on its own. Many of the suggested activities can reap tremendous benefits from partnering with specialized institutions to leverage high-quality results. The proposed Energy and Investment Outlook, for example, may be established in collaboration with organizations that have built up years of experience in gathering market research data and creating outlook documents for the global investment community. The CAREC program shall also consider new partnerships with renowned global centers of excellence to support the development of renewable energy in the region and other important areas targeted under this strategy.

THEME 2
Attracting Private Sector Investments across the Energy Value Chain

From 2016 to 2030, the CAREC region (excluding the People's Republic of China) will need estimated energy infrastructure investments amounting to $400 billion. Public sector investment has historically fallen too short of the necessary investment volume and is injected at a pace slower than required. Private investments will therefore have to fill the widening gap.

Creating the conditions for the emergence of a dynamic private sector—capable of diversifying exports, creating jobs, and strengthening resilience—is one of the most challenging and important concerns of the CAREC region, where the private sector remains underdeveloped and funding is limited. External private investment in the energy sector can relieve significant pressure on government budgets, besides improving the efficiency and viability of projects as they have to be profitable and competitive.

With this in mind, governments and development partners must create more space for the private sector by improving the business climate to strengthen interconnections, attract investments in energy efficiency and renewable energy, and stimulate competition in liberalized energy markets.

Support is also needed for small and medium-sized enterprises, to stimulate economic growth, combat poverty, and enhance energy development and regional linkages in the CAREC region.

Developing an Energy Investment Strategy for the CAREC Region

Many countries in the region are landlocked and nearest ports cannot be reached without cross-border connections. Therefore, energy infrastructure (and other infrastructure needed to bring in equipment and materials) must traverse borders and cannot be developed in isolation by individual countries. This makes investments in the CAREC region complex. Gradually aligning investment policy frameworks—so that investors face similar "rules of the game" across neighboring countries—can make a big difference in unlocking investment in cross-border projects.

To establish a solid base for policy makers to create effective enabling conditions for private investments in the energy sector, the CAREC program will develop a regional Energy Investment Strategy with suitable partner organizations. The strategy should benchmark investment policies, identify best practices, and provide policy makers with recommendations for enhancing investment policies as a whole to (i) make the investment environment more coherent and transparent; (ii) improve market access and healthy competition; (iii) protect investors' rights; and (iv) encourage responsible and inclusive investment.

Using Public–Private Partnerships

Using public–private partnerships (PPPs) as a financing mechanism is another way of helping the CAREC region attract much-needed private investment. PPPs will have an integral role in ensuring efficient risk sharing, identifying potentially successful projects, and establishing effective project monitoring schemes. They enable governments, which may lack resources and sophisticated knowledge to ensure the quality of large-scale projects, and distribute project risks among multiple parties.

Although the potential for PPPs remains largely unexplored, there have been promising developments in the region. Recently, many more countries have passed, or are in the process of passing, PPP laws and regulations. International financial institutions have supported these early developments and are taking on more active roles in building on these laws and regulations with early-stage transactions through advisory and, in many cases, financing support. PPPs are expected to multiply more rapidly across the region, especially in the energy sector.

Building on these instruments, the CAREC program shall develop a regional PPP template for interconnections based on best practices in the region and elsewhere to overcome the current barriers to private investment in cross-border energy infrastructure projects.

Promoting the CAREC Energy Investment Forum

The CAREC Energy Investment Forum, launched in 2016, has played a crucial role in showcasing business opportunities in the CAREC countries by bringing together key government officials, public and private financial institutions, project promoters, manufacturers and other stakeholders.

The forum shall continue to play a major role as the region's flagship platform for providing new knowledge about emerging technologies and trends as well as attracting investments to the region. It should not only feature matters that concern the established international and domestic investors but also actively promote business-to-business (B2B) exchanges, foster start-ups, and draw young entrepreneurs with new and innovative business ideas to the energy sector. The Energy Investment Forum shall also develop a series of concrete recommendations for government and business leaders after each Forum.

Creating a Business Advisory Council in the CAREC Region

Recognizing the value of bringing the voice of businesses into the regional energy discourse, the CAREC program will consider establishing a Central Asia Business Advisory Council (CABAC) to feed the views and needs of the private sector into regional policies. CABAC shall constitute a reality check in the process of creating an effective business climate in the region and articulating business priorities to encourage competition.

CABAC members shall consist of leaders from regional energy businesses, finance, and technology providers. They are expected to leverage their networks, knowledge, experience, and influence to:

- serve as sources of independent advice on energy business for the energy ministers in the CAREC region and the Energy Investment Forum;

- contribute significantly to the Energy Investment Strategy and the Energy and Investment Outlook for the CAREC region; and

- increase two-way engagement with the business, government, and the wider community.

CABAC shall achieve these objectives by convening, ideally around the time of the annual Energy Investment Forum and other relevant regional events, to support capacity development and investment promotion activities.

THEME 3
Empowering Women in Energy

Women remain underrepresented in the energy industry around the world, including the CAREC region. Today, women account for only around 20% of the global overall energy workforce,[6] and only 11% of top global oil and gas executives.[7] Hence, there is substantial room for opening up the energy industry to women and diversifying the sector's workforce. More women are needed at all levels in the organizational hierarchy, but especially at the senior executive level, to act as role models for their younger counterparts (Figure 7).

Women can multiply output and boost organizational performance in the energy sector. Strong support and collaboration from CAREC countries is therefore needed to ensure that gender equality is reached in a meaningful way across all sector activities.

Figure 7: Women in the Global Energy Workforce

20% of the global energy workforce are women

11% of top oil and gas executives are women

Sources: A. Brogan. 2019. How diversity boosts performance in oil and gas. EY. 23 January. https://www.ey.com/en_gl/oil-gas/how-diversity-boosts-performance-in-oil-and-gas; and Clean Energy Ministerial. Equal by 30: Gender Equality in the Clean Energy Sector. http://www.cleanenergyministerial.org/campaign-clean-energy-ministerial/equal-30.

Promoting Gender Equity and Added Value of Women in Energy

Raising awareness of the advantages of an inclusive and diverse workforce is crucial in achieving gender equality in the energy sector in the long term. Empowering women in energy shall bring about a change in mind-set among government and business leaders. More often than not, companies and governments choosing to bring women into their energy agenda have a more engaged and efficient staff, perform better, and maintain higher retention rates.

---

6   Clean Energy Ministerial. Equal by 30: Gender Equality in the Clean Energy Sector. http://www.cleanenergyministerial .org/campaign-clean-energy-ministerial/equal-30.

7   A. Brogan. 2019. How diversity boosts performance in oil and gas. EY. 23 January. https://www.ey.com/en_gl/ oil-gas/how-diversity-boosts-performance-in-oil-and-gas.

To showcase the added value contributed by women, the CAREC program shall introduce policies that will encourage women to participate in its energy events, and thus set a positive example. This approach has already produced positive results. Since 2016, women's participation at CAREC's Energy Sector Coordinating Committee (ESCC) meetings has gone up by 30% because of a new nomination policy for delegates. The CAREC program should keep pursuing this track and work toward achieving more far-reaching participation by women, not only in numbers but also in degree of active involvement in discussions.

Moreover, to increase awareness of the importance of gender equity, the CAREC program will organize a flagship Women in Energy Summit that gives prominence to the role of women in the regional energy sector. The summit should make women in energy more visible to a wider audience, showcase their talent, and allow for the creation of new role models. An excellent opportunity will also be provided to report on progress in achieving gender equity in the region and beyond.

Enhancing Employability of Women in Energy

Acknowledging the potential of women to participate formally in the energy sector as decision makers, and not merely as energy consumers, results in increased benefits. By creating a female talent pool, the CAREC program will ensure the rise of the next generation of women energy leaders and equip them with the tools they need to forge successful careers.

The CAREC program will establish a Women in Energy Program to increase the share of women in the energy sector by making them more employable and expanding their career opportunities in the sector. The program will establish a number of initiatives open only to women and offer targeted opportunities to enrich professional experience and sharpen skills. The aim is to boost the competitiveness of women on the job market. One way of achieving this goal is to set up a secondment program that allows women from the region to be employed for a defined period in the offices of international energy companies, public authorities, and international organizations. This will not only advance the career and leadership development of women but also encourage companies to rethink their business models and provide women with more significant roles in energy and associated sectors.

As part of the Women in Energy Program, the CAREC program will also enter into partnerships for higher education and training with elite universities worldwide. This move should create a competitive advantage for women in the region, particularly in its energy sector, as women with top-quality education can be expected to add fresh ideas and transform, inspire, and lead the regional energy industry to better results.

Fostering Regional Cooperation through Women Networks

Empowering women in the energy sector offers a valuable opportunity to unlock untapped productivity. For women to succeed on this journey, they will need support networks that provide guidance, inspiration, and the possibility of exchange among peers at all levels of society. The CAREC program will therefore establish a regional platform connecting urban female energy professionals as well as rural women's communities to enable them to discuss local and regional improvements in the energy sector and the contribution of women. The CAREC program should set up a suitable networking structure, keep it up to date, and ensure that women find the advice and support they seek. This network can include a mentoring program, with peers dispensing advice when needed, and initiatives connecting junior women with more senior counterparts. These connections should give rise to a new cross-border community of women in energy and contribute further to enhancing regional cooperation.

Women in energy.
Duty electrician writes data from the central control panel of a hydropower plant, Tajikistan.

V | INSTITUTIONAL FRAMEWORK IMPROVEMENTS

Achieving the 2030 objectives for the energy sector of the Central Asia Regional Economic Cooperation (CAREC) region is a complex task that requires a clear vision of how it can be effectively implemented. While a work plan with concrete actions and responsibilities is an essential part of this process,[8] optimal institutional structures that allow for a constructive working environment and high-quality results are equally important. This section proposes new and targeted work methods for delivering the objectives of the CAREC region's energy sector work up to 2030. The overall aim of this new institutional framework is to ensure better results, a more pronounced sense of ownership, and greater accountability in CAREC energy work.

A New Team Spirit for the CAREC Energy Program

The CAREC program must cater to the needs of the region. A strong sense of community is required to identify, develop, and produce beneficial results for all parties involved. A regional community whose members support one another and strive toward a common purpose is crucial in building a dynamic and high-performing network. Over the next decade, the CAREC energy program shall thus become a platform operated by its members for its members. It should evolve further into a unique platform that brings energy experts, companies, governments, and international financial institutions together to formulate plans and actions for emerging challenges.

Bringing people to the forefront of the CAREC energy program requires building a community with names, faces and personalities. Hence, the program must provide appropriate structures for CAREC members to come under the spotlight while taking on responsibility and leadership for specific tasks. Who is who in the CAREC energy program must be well known to all members of the community to increase ownership and accountability for results.

Overall, a new and smart approach shall be applied to achieve the objectives of the CAREC Energy Strategy 2030. A new culture of teamwork shall allow the sharing of workload and more focused discussions about the individual focus areas. Teams shall lead the community in developing a fresh identity and a new team spirit. This, in turn, shall motivate members to work toward a common goal and build closer ties.

Ultimately, the CAREC Energy Strategy 2030 is owned by its members. Its outcomes will be only as strong as its actors. For this reason, members must be provided with optimal working conditions to achieve maximum impact for the energy sector in the region and take regional cooperation to new heights in the coming years.

---

[8] Covered in Section VI of this strategy document.

Delivering More and Better

The CAREC program's main discussion platform for energy issues is the Energy Sector Coordinating Committee (ESCC). The ESCC's performance is critical as it determines the quality and level of output of the CAREC program's energy sector work. Its effective functioning is thus essential for the creation of lasting results and value added for the regional energy community.

So far, the ESCC has been meeting around twice a year to present regional sector updates. As it looks ahead to 2030, given the scope of the new regional priorities, the committee will need to carry out more in-depth discussions and effective division of tasks. To this end, appropriate substructures that allow individual subject areas to be tackled in a more targeted fashion should be put in place to increase the productivity of this uniform body. The ESCC shall thus be transformed into a more output-oriented organization with permanent subgroups, each delivering assigned strategic results.

The new ESCC structure shall mirror the strategy pillars by putting in place work streams for each of the main priority areas (Figure 8):

• Work Stream 1 (WS 1): Infrastructure Connectivity and Energy Security;

• Work Stream 2 (WS 2): Policy Reform and Liberalization; and

• Work Stream 3 (WS 3): Energy Efficiency and Diversification of the Energy Mix.

These work streams directly cover the scope of strategy pillars 1, 2, and 3 and shall focus on their implementation. Progress and results shall be reported regularly to the ESCC for further guidance and adoption of final results.

The crosscutting themes shall be covered in similar fashion. Individual task forces for each of the crosscutting themes shall be established and convened on demand:

• Task Force A (TF-A) on Knowledge Products, Partnerships, and People's Networks;

• Task Force B (TF-B) on Private Sector Enhancement; and

• Task Force C (TF-C) on Women Empowerment.

While members of the task forces must have their own discussions, they must also be represented in each of the working groups to ensure the full transparency of crosscutting actions to all members.

Both work streams and task forces shall establish work programs with concrete deliverables in accordance with the overall work plan up to 2030.

Working groups and task forces shall each be led by a chair and a co-chair, from two different member countries, if possible. Chairs and co-chairs are equal in rank and shall be responsible for steering their groups and producing the anticipated results. They must take part in ESCC meetings and report regularly on the progress and results achieved by their groups.

Figure 8: New Institutional Framework for the CAREC Energy Program

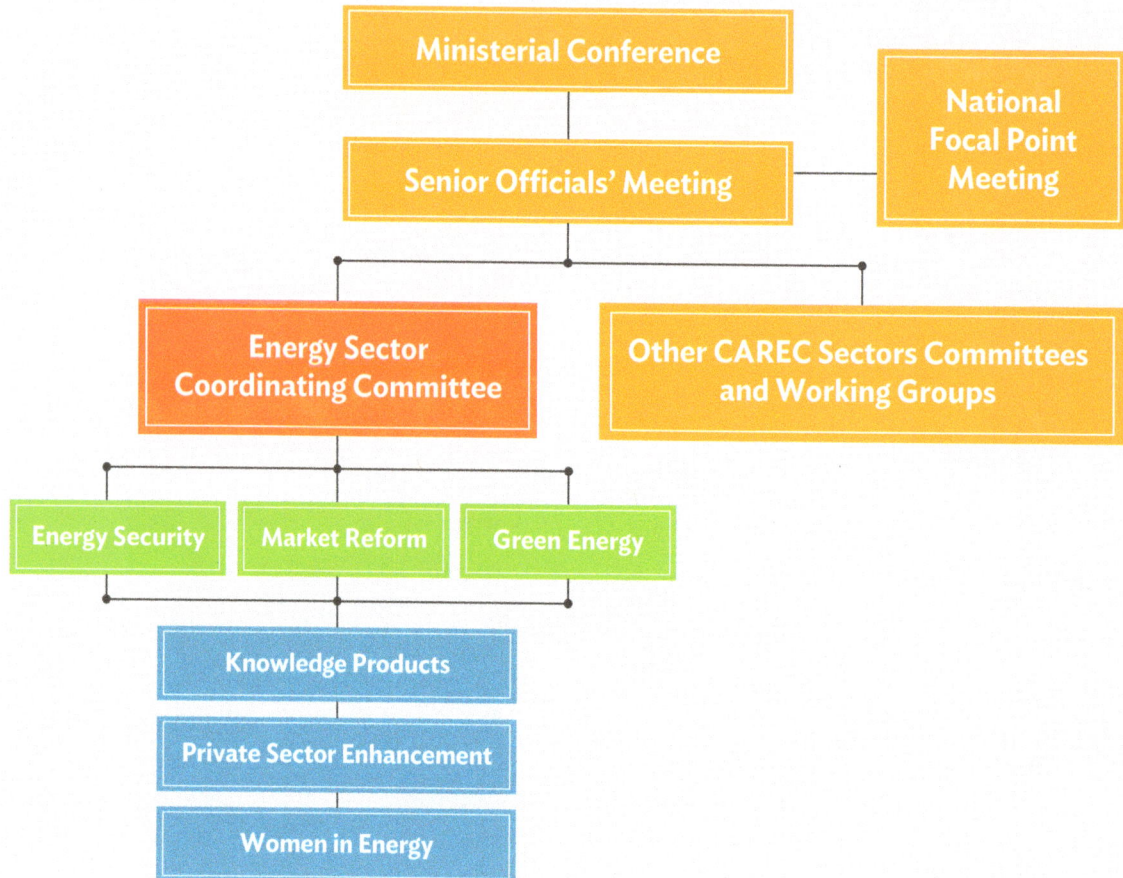

```
                    ┌─────────────────────────┐
                    │  Ministerial Conference │
                    └─────────────────────────┘
                                 │                    ┌──────────────┐
                    ┌─────────────────────────┐       │   National   │
                    │ Senior Officials' Meeting├───────┤ Focal Point  │
                    └─────────────────────────┘       │   Meeting    │
                                 │                    └──────────────┘
         ┌───────────────────────────────┐   ┌──────────────────────────────┐
         │      Energy Sector            │   │  Other CAREC Sectors         │
         │ Coordinating Committee        │   │  Committees and Working      │
         │                               │   │  Groups                      │
         └───────────────────────────────┘   └──────────────────────────────┘
         ┌───────────┐ ┌─────────────┐ ┌──────────────┐
         │  Energy   │ │   Market    │ │ Green Energy │
         │ Security  │ │   Reform    │ │              │
         └───────────┘ └─────────────┘ └──────────────┘
              ┌───────────────────────────┐
              │   Knowledge Products      │
              └───────────────────────────┘
              ┌───────────────────────────┐
              │ Private Sector Enhancement│
              └───────────────────────────┘
              ┌───────────────────────────┐
              │     Women in Energy       │
              └───────────────────────────┘
```

CAREC = Central Asia Regional Economic Cooperation.
Source: ADB.

The new ESCC structure shall give CAREC members full ownership, responsibility, and control of the 2030 work plan and its deliverables.

Facilitating Output through Smart Collaboration Platforms

For regional teamwork to be effective, modern technology and state-of-the-art collaboration tools shall be used to facilitate working across borders. A web-based platform shall serve as the main gateway for the working group and task force members in implementing the Energy Strategy 2030, gaining access to work documents, planning meetings, and sharing ideas.

Gas supply in rural Central Asia.
Gas pipelines cross the Almaty–Bishkek
highway to provide natural gas to villages
along the highway.

The following work plan has been established for the implementation of the CAREC Energy Strategy 2030. It contains the actions to be delivered for each of the strategy pillars and crosscutting themes, and sets out specific performance indicators for tracking progress (Tables 1–6). The work plan also determines which subgroups shall be responsible for the delivery of the actions. The final shape of results will be worked out together with the CAREC members, as part of the strategy implementation process, to best suit individual and regional needs.

Pillar 1 | Better Energy Security through Regional Interconnections

Table 1: Actions to Be Delivered by Work Stream 1—Infrastructure Connectivity and Energy Security

Action	Description	Performance Indicators
Realize TUTAP, TAP, and CASA-1000 electricity interconnection projects	The CAREC region's flagship power transmission line projects connecting Central and South Asia are at various stages of implementation and shall go into operation within the strategy period.	• TUTAP, TAP, and CASA-1000 projects in operation
Advance TAPI gas pipeline project	Ongoing negotiations for possible modalities to realize the TAPI gas pipeline shall be accelerated.	• Dialogue on the implementation of TAPI project intensified
Facilitate cooperation among regional transmission system operators	Growing electricity and gas interconnections require increased cooperation among transmission system operators. The establishment of a corresponding platform for regional network development planning, identification of projects of common interest, and information sharing shall be facilitated by this activity.	• Central Asia Transmission Cooperation Association (CATCA) concept developed
Identify new cross-border infrastructure projects	New cross-border gas and electricity links shall be identified to increase energy security in the region.	• New regional gas and/or power interconnections identified

CAREC = Central Asia Regional Economic Cooperation Program, CASA = Central Asia–South Asia, TAP = Turkmenistan–Afghanistan–Pakistan, TAPI = Turkmenistan–Afghanistan–Pakistan–India, TUTAP = Turkmenistan–Uzbekistan–Tajikistan–Afghanistan–Pakistan.
Source: ADB.

Pillar 2 | Scaled-Up Investments through Market-Oriented Reforms

Table 2: Actions to Be Delivered by Work Stream 2—
Policy Reform and Liberalization

Action	Description	Performance Indicators
Build capacity for unbundling models and liberalization of energy markets	This activity shall support policy makers in making informed decisions when embarking on unbundling and market liberalization reforms.	• CAREC Energy Reform Atlas (containing access to practical handbooks and database with study materials) established
Advise on tariff-setting principles and methods of introducing tariff reform	Financial health of network companies is critical to ensuring high quality of service. This activity shall shed light on tariff design options and ways of implementing gradual tariff reform.	• Handbook on Tariff Principles and Reform Options published • Capacity building workshops held
Develop options for social protection measures for vulnerable energy consumers	This activity shall assist in elaborating options for social protection measures to accompany tariff reform, and shall include the development of a definition for "vulnerable consumers" to assist policy makers in identifying the appropriate target group.	• CAREC Guide to Social Protection and Energy Sector Reform published
Share international best practices in legal enforcement of sector reform	Sector reform requires adapting existing energy laws and creating new laws. This activity is aimed at discussing successful examples of relevant sector laws and methods of law enforcement.	• Capacity building workshops held

CAREC = Central Asia Regional Economic Cooperation.
Source: ADB.

Pillar 3 | Enhancing Sustainability by Greening the Regional Energy System

Table 3: Actions to Be Delivered by Work Stream 3—
Energy Efficiency and Diversification of the Energy Mix

Action	Description	Performance Indicators
Establish a joint platform for mobilizing sources of funding for emission reduction projects	For emission reduction projects to be realized, a dedicated platform shall be established for identifying and securing funding for priority projects.	• CAREC Green Energy Alliance established
Identify suitable energy efficiency measures and track progress	Energy efficiency is a powerful tool for reducing emissions. The aim of this activity is to disseminate practical skills in implementing efficiency measures and establish a regional benchmarking tool for comparing progress and rewarding high performers.	• CAREC Energy Efficiency Week held • Regional Energy Efficiency Scorecard developed • Capacity building workshops held

continued next page

Table 3: *Continued*

Action	Description	Performance Indicators
Create public awareness of energy efficiency	This activity shall enhance public awareness of energy efficiency and empower consumers to engage in more conscious use of energy.	• CAREC Consumer Leaflet on Energy Saving prepared • Energy efficiency radio or TV commercial developed • Handbook for organizing consumer awareness campaigns published
Support in the development of renewable energy and the diversification of the energy mix	This activity is aimed at supporting CAREC members in adding renewable energy to their energy mix by providing practical guidance on the necessary pre-requisites.	• Workshop on pros and cons of different renewable incentive schemes held • Coping mechanisms for renewable energy intermittency developed • Manual on legislative requirements for the integration of renewable energy prepared • Role of gas as a transition fuel discussed

CAREC = Central Asia Regional Economic Cooperation.
Source: ADB.

Cross Cutting Themes

Theme 1 | Building Knowledge and Forming Partnerships

Table 4: Actions to Be Delivered by Task Force A—
Knowledge Products, Partnerships, and People's Networks

Action	Description	Performance Indicators
Develop CAREC Energy Outlook and Investment Report	This activity is aimed at providing investors and other relevant stakeholders with reliable regional data to make investment and policy decisions.	• CAREC Energy Outlook and Investment Report published
Establish relevant partnerships to support the implementation of the three strategy pillars	All three strategy pillars include capacity building, institution-building, and training initiatives, which shall be implemented in collaboration with experienced partner organizations.	• Partnerships with global centers of excellence in the fields covered by the strategy established
Promote networking and skills development of next-generation energy professionals	The CAREC program shall facilitate cross-regional networking and skills development of next-generation energy professionals to allow a natural sense for regional cooperation to emerge among the target group.	• CAREC Young Energy Leaders initiative established

CAREC = Central Asia Regional Economic Cooperation.
Source: ADB.

Theme 2 | Attracting Private Sector Investments across the Energy Value Chain

Table 5: Actions to Be Delivered by Task Force B—
Private Sector Enhancement

Action	Description	Performance Indicators
Prepare a regional investment strategy	A regional investment strategy containing recommendations for improved enabling conditions for private investments in the CAREC region shall be developed.	• CAREC Energy Investment Strategy developed
Hold annual CAREC Energy Investment Forum	The annual Energy Investment Forum shall continue to attract investors to the region and foresee a dedicated space for B2B meetings.	• CAREC Energy Investment Forum held yearly
Provide practical support to investors to enhance their business operations in the region	This activity shall create improved enabling conditions for private investors operating in the region.	• CAREC Business Advisory Council created to identify needs of private investors

B2B = business-to-business, CAREC = Central Asia Regional Economic Cooperation.

Source: ADB.

Theme 3 | Empowering Women in Energy

Table 6: Actions to Be Delivered by Task Force C—
Women's Empowerment

Action	Description	Performance Indicators
Establish CAREC Women in Energy Program	This activity shall provide women in the CAREC region with the necessary tools to boost their career, build a regional network, and become more visible in the region's energy sector.	• Women in Energy Summit organized • Secondment program for women in energy established • Educational scholarship program for women in energy facilitated

CAREC = Central Asia Regional Economic Cooperation.

Source: ADB.

Treasures of nature.
The Pamir Mountain Range is known
as "the roof of the world," Tajikistan.

VII | RESULTS FRAMEWORK

Table 7: CAREC Energy Strategy 2030 Program Results Framework

	Common Borders. Common Solutions. Common Energy Future.					
Impact ↑	Better Energy Security through Regional Interconnections	Scaled-Up Investments through Market-Oriented Reforms	Enhanced Sustainability through Greening the Regional Energy System	New Knowledge Products and Partnerships	Enhanced Role for the Private Sector	Empowered Women in Energy
Outcome	Increased energy security as a result of cross-border collaboration	Higher quality of service without disadvantage to the most vulnerable	Participation of the CAREC region in global efforts to mitigate climate change	More high-quality data on regional energy markets and engaged young professionals	Increased private sector confidence in investing in the CAREC region	Increased women's participation in the energy sector
Output ↑	Regional platform for transmission system operators is created New projects of common regional interest are identified	Capacity for tariff setting, protection of vulnerable consumers, and subsidy phaseout are strengthened	Regional financing vehicle for clean energy is established Incentive mechanism for more efficient energy use is created Capacity for renewable energy integration is enhanced	New energy publication with regional energy market data is released Regional cooperation is intensified through a platform for young energy professionals	Improved enabling conditions for private investments are created B2B platforms are made available to investment community Support networks for private sector are created	Opportunities for women to improve qualifications and employability are created
CAREC Interventions ↑	Establish concept for a Central Asia Transmission Cooperation Association (CATCA) Support priority implementation of projects of common regional interest	Establish a virtual CAREC Energy Reform Atlas Release a Handbook on Tariff Setting and Reform Options Develop a Guide to Social Protection and Energy Sector Reform	Establish a CAREC Green Energy Alliance Develop a CAREC Energy Efficiency Scorecard Build capacity for renewable energy integration and uptake Organize public awareness campaigns for energy efficiency	Develop a CAREC Energy Outlook and Investment Report Launch Young Energy Leaders Initiative	Develop a CAREC Energy Investment Strategy Organize regular Energy Investment Forums to provide possibility of B2B exchange	Establish a Women in Energy Program
Driving principles ↑	Energy security Long-term and enduring regional relationships Economic gains for all	Financial sustainability of the energy sector Affordable and reliable service quality to consumers	Emission reductions through energy efficiency and clean energy generation and use Attraction of private finance Diversification of economy and energy mix	Transparent energy markets Higher-quality results through specialized partnerships Fresh ideas and innovation from young professionals	Attraction of private investment to cover the region's energy investment needs	Realization of proven advantages of a diverse and inclusive workforce CAREC program leading by example and acting as champion of increased women's participation in the energy sector

B2B = business-to-business, CAREC = Central Asia Regional Economic Cooperation.
Source: ADB.

www.ingramcontent.com/pod-product-compliance
Lightning Source LLC
Chambersburg PA
CBHW040147200326
41519CB00035B/7618